T3-BNI-690

The
REPRODUCTION
OF PROFILES

DISCARDED

By Rosmarie Waldrop

The Aggressive Ways of the Casual Stranger (Random House)
The Road Is Everywhere or Stop This Body (Open Places)
When They Have Senses (Burning Deck)
Nothing Has Changed (Awede)
Differences for Four Hands (Singing Horse Press)
Streets Enough to Welcome Snow (Station Hill)
The Hanky of Pippin's Daughter (Station Hill)

Translations:

The Book of Questions by Edmond Jabès (Wesleyan University Press)
The Vienna Group: Six Major Austrian Poets (with Harriett Watts, Station Hill)
Archeology of the Mother by Alain Veinstein (with Tod Kabza, Spectacular Diseases)

ROSMARIE WALDROP

The

REPRODUCTION
OF PROFILES

A NEW DIRECTIONS BOOK

Copyright © 1984, 1985, 1986, 1987 by Rosmarie Waldrop

All rights reserved. Except for brief passages quoted in a newspaper, maga-
zine, radio, or television review, no part of this book may be reproduced
in any form or by any means, electronic or mechanical, including photo-
copying and recording, or by any information storage or retrieval system,
without permission in writing from the Publisher.

Grateful acknowledgment is made to the editors and publishers of maga-
zines and journals in which sections of this book first appeared: *Acts*,
Brooklyn Review, *Central Park*, *Cold Water Business*, *Conjunctions*, *Constant Red/
Mingled Damask*, *Feminist Studies*, *Gallery Works*, *Grosseteste Review*,
HOW(ever), *In Folio*, *New American Writing*, *Ninth Decade*, *O·blēk*, *Oovrah*,
Open Places, *Reality Studios*, *Sink*, *Southpaw*, *Temblor*, and *Tyuonyi*.

Manufactured in the United States of America
First published clothbound and as New Directions Paperbook 648 in 1987
Published simultaneously in Canada by Penguin Books Canada Limited

Library of Congress Cataloging-in-Publication Data

Waldrop, Rosmarie.
    The reproduction of profiles.
    (A New Directions Book)
    I. Title.
PS3573.A4234R45   1987   811'.54   87-11038
ISBN 0-8112-1044-8
ISBN 0-8112-1045-6 (pbk.)

New Directions Books are published for James Laughlin
by New Directions Publishing Corporation
80 Eighth Avenue, New York 10011

for Keith Waldrop

Contents

PART ONE

THE
REPRODUCTION
OF PROFILES

I

Facts

I had inferred from pictures that the world was real and therefore paused, for who knows what will happen if we talk truth while climbing the stairs. In fact, I was afraid of following the picture to where it reaches right out into reality, laid against it like a ruler. I thought I would die if my name didn't touch me, or only with its very end, leaving the inside open to so many feelers like chance rain pouring down from the clouds. You laughed and told everybody that I had mistaken the Tower of Babel for Noah in his Drunkenness.

I didn't want to take this street which would lead me back home, by my own cold hand, or your advice to find some other man to hold me because studying one headache would not solve the problem of sensation. All this time, I was trying to think, but the river and the bank fused into common darkness, and words took on meanings that made them hard to use in daylight. I believed entropy meant hugging my legs close to my body so that the shadow of the bridge over the Seekonk could be written into the hub of its abandoned swivel.

The proportion of accident in my picture of the world falls with the rain. Sometimes, at night, diluted air. You told me that the poorer houses down by the river still mark the level of the flood, but the world divides into facts like surprised wanderers disheveled by a sudden wind. When you stopped preparing quotes from the ancient misogynists it was clear that you would soon forget my street.

I had already studied mathematics, a mad kind of horizontal reasoning like a landscape that exists entirely on its own, when it is more natural to lie in the grass and make love, glistening, the whole length of the river. Because small, noisy waves, as from strenuous walking, pounded in my ears, I stopped my bleak Saturday, while a great many dry leaves dropped from the sycamore. This possibility must have been in color from the beginning.

Flooding with impulse refracts the body and does not equal. Duck wings opened, jeweled, ablaze in oblique flight. Though a speck in the visual field must have some color, it need not be red. Or beautiful. A mountain throwing its shadow over so much nakedness, or a cloud lighting its edges on the sun, it drowned my breath more deeply, and things lost their simple lines to possibility. Like old idols, you said, which we no longer adore and throw into the current to drift where they still

II

Thinkable Pictures

Only in connection with a body does a shadow make sense. I called mine a dog, the way it ran ahead of me in the dust, breathing rapidly and sticking its small head out in front—though there are intervals where the light stands still, and the air does not resist. Abandoned in my body, the memory of houses at a certain distance, their roofs, and their chimneys for the dark to flow down in arbitrary conventions. This is why you don't like me to get drunk. I fall asleep in the street, without even a shadow to lie on, and crowds gather, afraid of being disappointed.

Was it the jokes you told? Our bodies fitted one another like the links of a chain resplendent with cymbals and xylophones. Because form is the possibility of structure, you hoped there were people watching. My desire was more like a sailor's rolling gait, as if shifting my weight from one side to the other were a matter heavy with consequence. The salt reached saturation. You said wet was wet, without following the river farther than this sentence or looking at negative facts, their nonexistent mouths twisted for explanation.

I was not sure I had understood. I was naked enough to disappear in the shop windows. Your weight on me sank through my bones, and I didn't know where I had lost my body—as if it had no vowels, as if the construction were faulty, the mesh too coarse—when you felt a sneeze coming on and fumbled for your handkerchief. I traced the law of sufficient reason down your spine. Your skin was delicate, like a retracted confession.

Everything that can be thought at all, you said, can be thought over. When I asked if you were referring to nuclear arms, genetic engineering, or marriage, you hastily closed the window. I had seen you, in the park, push a banana peel off the sandal of Constance Witherby's statue and recite with large gestures: a poem? a funeral oration? I was not musician enough to read this score, not with the wind blowing your hair against the approach of winter, though if the swallows had stopped circling high in the solid blue, my breath would already have failed me. Sharp smell of the sea, of fish rocking in the surf. And already clouds. You said it might be different if we were able to stand outside logic. I knew by this you meant: barefoot.

For years already, the countryside had been in competition with my thoughts, not like weather moving through the lungs, but pulling. It was beautiful. It wanted to be looked at. And my attention swayed like a poplar in the wind. Nevertheless I had learned to substitute "freedom of the will" for not knowing which future would undress me and pretended to be happy to hold my breath. So I agreed, you said, that there were languages to be admired rather than understood, and that my smiles shot dubious appeals to the passage of time though we knew the river flowed a few yards off whether we tried to cross it or not. The trees rocked gently in the fine mist. Mostly, I had to admit, I live in the subjunctive, unable to find a foothold. If objects were doors, I might be drawn north to forget them.

Waves can be resolved into a statement about unalterable form. It goes without saying they lap at your chin while you are still describing the dangers of dry land, after having loved it so long. The gulls stood still, though the light fell on their strained bodies. They could not be proved within the substance of the world, you said. So different their flight. I was happy to discover cause, the better to ignore effects. The clouds drowned silently in their reflection, pulling water down with them.

We were approaching winter like an object which cannot be put between words. Behavior became simpler since we had dislocated our memories. Still, much was. A little confusion in the propositions will allow for this. Or truth can be so strenuous it makes you lean against the window frame. I thought of breathing deeply to find Venus reflected in the river. Then I would know if standing beside you leaves my lips dry. But I was really dissecting your name by means of definitions which would point the way to the missing copula where I could see the sky. Though the clouds could be uttered in a variety of tones, the stars formed constellations analyzed completely. You cried for the moon, which had started to wane in agreement with constant and variable. What this silver sliver failed to reveal, its expression between my thighs would clarify.

There were obstacles, no doubt about it. Take the huge plain rising against the sky and which we must cross while all we have done in our sleep falls away. Then stipulate singing. If I fail to deposit a coin, everyday language produces the most fundamental confusions, but what pleasure in getting lost if it is unavoidable? Fear, possibly. A vague distance instead of horizon. You were wondering if the same path could be taken by two people. And if any grass had grown on the runway since J. F. Kennedy's take-off.

III

Feverish Propositions

You told me, if something is not used it is meaningless, and took my temperature which I had thought to save for a more difficult day. In the mirror, every night, the same face, a bit more threadbare, a dress worn too long. The moon was out in the cold, along with the restless, dissatisfied wind that seemed to change the location of the sycamores. I expected reproaches because I had mentioned the word love, but you only accused me of stealing your pencil, and sadness disappeared with sense. You made a ceremony out of holding your head in your hands because, you said, it could not be contained in itself.

If we could just go on walking through these woods and let the pine branches brush our faces, living would still make beads of sweat on your forehead, but you wouldn't have to worry about what you call my exhibitionism. All you liked about trees was the way the light came through the leaves in sheets of precise, parallel rays, like slant rain. This may be an incomplete explanation of our relation, but we've always feared the dark inside the body. You agree there could be no seduction if the structures of propositions did not stand in a physical relation, so that we could get from one to the other. Even so, not every moment of happiness is to hang one's clothes on.

I might have known you wouldn't talk to me. But to claim you just didn't want to disguise your thoughts! We've walked along this road before, I said, though perhaps in heavier coats not designed to reveal the form of the body. Later, the moon came out and threw the shadows of branches across the street where they remained, broken. Feverishly you examined the tacit conventions on which conversation depends. I sighed as one does at night, looking down into the river. I wondered if by throwing myself in I could penetrate to the essence of its character, or should I wait for you to stab me as you had practiced in your dream? You said this question, like most philosophical problems, arose from failing to understand the tale of the two youths, two horses, and two lilies. You could prove to me that the deepest rivers are, in fact, no rivers at all.

From this observation we turned to consider passion. Looking at the glints of light on the water, you tried to make me tell you not to risk the excitement—to recommend cold baths. The lack of certainty, of direction, of duration, was its own argument, unlike going into a bar to get drunk and getting drunk. Your face was alternately hot and cold, as if translating one language into another—gusts from the storm in your heart, the pink ribbon in your pocket. Its actual color turned out to be unimportant, but its presence disclosed something essential about membranes. You said there was still time, you could still break it off, go abroad, make a movie. I said (politely, I thought) this wouldn't help you. You'd have to kill yourself.

Tearing your shirt open, you drew my attention to three dogs in a knot. This served to show how something general can be recorded in unpedigreed notation. I pointed to a bench by a willow, from which we could see the gas tanks across the river, because I thought a bench was a simple possibility: one could sit on it. The black hulks of the tanks began to sharpen in the cold dawn light, though when you leaned against the railing I could smell your hair, which ended in a clean round line on your neck, as was the fashion that year. I had always resented how nimble your neck became whenever you met a woman, regardless of rain falling outside or other calamities. Now, at least, you hunched your shoulders against the shadow of doubt.

This time of day, hesitation can mean tottering on the edge, just before the water breaks into the steep rush and spray of the fall. What could I do but turn with the current and get choked by my inner speed? You tried to breathe against the acceleration, waiting for the air to consent. All the while, we behaved as if this search for a pace were useful, like reaching for a plank or wearing rain coats. I was afraid we would die before we could make a statement, but you said that language presupposed meaning, which would be swallowed by the roar of the waterfall.

Toward morning, walking along the river, you tossed simple objects into the air which was indifferent around us, though it moved off a little, and again as you put your hand back in your pocket to test the degree of hardness. Everything else remained the same. This is why, you said, there was no fiction.

IV

If Words Are Signs

In order to understand the nature of language you began to paint, thinking that the logic of reference would become evident once you could settle the quarrels of point, line, and color. I was distracted from sliding words along the scales of significance by smoke on my margin of breath. I waited for the flame, the passage from eye to world. At dawn, you crawled into bed, exhausted, warning me against drawing inferences across blind canvas. I ventured that a line might represent a tower that would reach the sky, or, on the other hand, rain falling. You replied that the world was already taking up too much space.

Two sailors throwing dice on the quay will not make a monument, but there you sat reading a paper in its shadow. You said once we had a language in which everything was alright, everything would be alright, and your body looked beautiful while a fisherman tied his boat to a post, looping his rope through the metal rings without getting entangled in problems of representation or reflection. Nobody looked at you except for the water which, though it has no shape, is heavy with mirroring that of others. These images, however, are hard to get hold of, sunk as they are at the bottom of the alphabet.

The shifting use of the word "home" corresponds to dim light. A clothesline with pieces of torn underwear, reflected in a puddle. I stood in the yard looking for clues, a cat on the banister, the tick of a clock from years ago. Even as a memory the house was dark. Impossible to distinguish forms on the tip of my tongue. Cold, the closed door, the steps leading up to it. As I was listening at a distance I could only hear the distance. It pulsed in my ears, and the longer I waited the more it. This brings us again, you said, to the vexed question whether desires are internal or more like foreign countries.

At first sight, it did not look like a picture of your body. Any more than the fog rolling in from the sea, covering and uncovering the surface of the river, seemed an extreme. I made excuses for your hesitation because I thought you wanted to contain everything, unimpaired by spelling errors. Then I saw you were trying to lean against the weight of missing words, a wall at the end of the world. But I knew, though it tired me to imagine even a fraction of the distance, that it continued at least as far as one can run from danger, where two women had been washed up on a delay. Neither words nor the rigor of sentences, you said, could stem the steady acceleration of the past.

As the streets were empty in the early morning, I had made the spaces between words broad enough for a smile which could reflect off the enamel tower clock. Being late is one of my essential properties. Unthinkable that I should not possess it, and not even on vacation do I deprive myself of its advantages. Nevertheless I cannot recall a time when I did not try to hide this by changing the shape of my mouth and appearing breathless. The sky was shading from hesitant to harsh, which was not bound to correspond to any one color or tableau vivant. The climate is rainy, no doubt about it, and ready to draw its curtain over my clauses and conjunctions. But what if I had made the spaces too wide to reach the next word and the silence

The Seekonk was losing its shape, flooding, forming a sea of its own by way of experiment. You thought if we stopped making pictures, the word "I" would be terrifying with vagueness, and a massive silence would spread through all the algae and ruined crops. We'd stagnate in puddles, without elegance or variety, before drying into thin sleep. Nothing could drink its fill from our lacking sentences, because if a river has no movement no tales are left dreaming in it. All this time, I tried to describe a blot of ink on white paper by stating for each point on the sheet whether it was black or white.

I had tried to find ways of escaping my daily chores, but then read the river from one bank to the other and took longer. The words must already have sense. They could not acquire it through reflection, no matter how strenuous the pigment. Indeed, the light came on inside, to reveal the waters of childhood carrying their junks and packets. If I could only accept similes, you began, but I interrupted with a question about your body of doctrine. That, you said, would take rhythm and logic in afternoon rotation. You preferred to speak of years that pour down like whiskey.

The fog was not dense enough to hide what I didn't want to see, nor did analysis resolve our inner similarities. When you took the knife out of your pocket and stuck it into your upper arm you did not tell me that, if the laws of nature do not explain the world, they still continue its spine. There was no wind, the branches motionless around the bench, a dark scaffolding. A few drops of blood oozed from your wound. I began to suck it, thinking that, because language is part of the human organism, a life could end as an abrupt, violent sentence, or be drawn out with economy into fall and winter, no less complicated than a set of open parentheses from a wrong turn to the shock of understanding our own desires.

V

Successive Applications

The party began to break up though you were still looking for a point of view by examining thoughts for possible sexual characteristics. People assured one another it had been a nice evasion, when the floor that slanted downward in the mirror was suddenly pulled up to the surface on which you stood, disheveled and exhausted. I understood your desire to communicate, but stepped over it because I was thirsty. I had meant to tell you that it is improper to speak of sex to a person drinking cognac, but not even sober could I have handed you a sequence of missing links.

We can now talk about formal sex in the same sense that we speak of formal concepts, you roared with violent gallantry, but this woman, my God! She was showing us downstairs. At the door, she pressed her body against yours and pressed and pressed until I put a quarter in her hand. Then she covered her mouth as if in fright or in order to protect the wet impact of your lips or, again, to keep a cry from rising into the air on large, trembling wings. I introduce this metaphor in order to get to the source of your confusion between formal sex and sex proper, which has looped the whole of traditional philosophy to the moment, toward the end of day, when the equator embraces the torrid zone.

I could see sleepiness round your skinny limbs, so many dreams clinging to the bone, as I knocked on wood, and the door opened onto the cold. When snow falls under this category we must advance carefully, with small steps. But a large moon now lit up the sky which seemed more navigable for its occasional cloud. Twice I reached out to take your hand, but as you paid no attention I put mine back in my pocket until further need. I moved on to saying your name, which shows that it's easier to label a cold front than to predict the wedding. You said, the total number of objects was already breathless with growing pains.

You said I should have let her kiss me, too. But now it was too late, my hand cold as an absence we don't doubt because our understanding so palpably depends on it. I knew all I could do was go home and smother my thoughts under my blanket as one does when one sleeps alone. Then dreams carry the body from the shadow line into a brief splendor, and China could not be less remote. I thought I would call you a joker for the hold it would take in your backbone and tease you on. You said a dream was a truth like any other, but to outstrip an adversary with tautologies which, as we know, vanish inside themselves, was an extreme case of provoking phantoms.

Nail clippings have, thanks to Sigmund Freud, become mysterious again, you continued, whereas a proposition flaunts every logical scratch that follows from it. I felt sleepy, no doubt because I have a long past and don't speak foreign languages. The shadows made the ground sink a calculable distance. The dark helps, I said. Open your hands, and there is time in the depth of memory—a mirror with ambition—which assures both interval and continuity by letting slow breathing rise to the surface. Making a smooth stone skim across the water, you replied that the relation of a policeman to a crime about to take place could not be inferred from polluted air. Causality was only one way of losing the world.

You were leaning over the parapet. Night allowed us to see as far back as Roger Williams' landing and the anger which scraped faces off the dead. So much unsolved memory under the bridge. My thoughts began to share the darkness of the river, though we were miles from the nearest reactor. Tips of grass stuck up through the snow. You suddenly smiled into the latency between us and said that her breasts were very white. I felt I ought either to wash this episode out of my centerfold or kiss you for having so little use for me.

Snowflakes floated to the ground with infinite caution, accumulating a silence into which one could not introduce primitive discrepancies. I listened to the muffled thud of our steps and wondered why I could not keep pace with you, even though I clearly saw your feet. You said that "one plus one at midnight equals two" was as nonsensical as a nephrite worn as a charm against kidney stones or a day without birds. A woman opened her window and overlooked the difference between the sexes while you complained that the Milky Way, albeit invisible at the moment, was caused by another long proposition, not unlike amnesia.

It was easier to walk in the snow if I made swimming motions, so I let the moonlight frisk my body which it threatened to reduce to a mere projection. Your body seemed to flicker as you told me how the insides of her thighs lay peacefully in the mirror with no thought for consequences. Suppose, I said patiently, kicking at a snowdrift, I am given *all* the details at once. Then I could construct all possible stories out of them, and that would be the end of it. Annoyed, you bit your native tongue. But I knew well enough that if one leaves things alone they get less clear by themselves.

You were walking ahead, humming Berlioz to keep me from introducing more conditional clauses, from which anything might follow. I predicted your approximate indifference, should I simply take a side street home. Instead, I took a sharp breath and held it, letting the cold melt in my lungs like distance. It brought you skidding toward me on the snow. Winking, you asked: if a heart was infinitely complex, so that its every desire was for infinitely many stiffenings with infinitely many terrors, rather than for risking that a single scream ravage all memory, was such a heart not implausible with ambition and therefore bound to succeed? I said a complete description of the world was given by listing one's fears and then listing one's fulcrums, but you were mocking from the back of a mirror, the silver of multiplication.

Then you came out with how you had jumped her and, trembling with courage, ridden up the hill into the interior, a landscape with something unfinished about it. I was tempted to register doubts as your description progressed, but the wind died down. Cold, at this hour. The moon sank among the clouds as into a lake. A large bird, an owl perhaps, swooped from a tree, and we expected to see it fly up again, its silhouette bulging with prey. Instead, we saw another bird swoop down and, after a while, another. Regardless whether it was several owls or the same, you said, they could be arranged on the road and treated as outlaws of probability. But I hesitated, for fear of not encountering.

It is clear that distance devours the variables and leaves us with all propositions saying the same thing, but with such force that the desire takes us out of body. Tell me that she is beautiful, you demanded, even though you knew that I had always been pleased to lead you astray. A name, I said, cannot go from mouth to mouth, a clear mirror unclouded by breath. Remember that nightingales sing only in the upper pay scales. And we can't logically correlate a fact with a soul, even if fiction sustains the tone of our muscles. Your lips trembled slightly as you said that logic could take care of itself.

PART TWO

INSERTING THE MIRROR

1 To explore the nature of rain I opened the door because inside the workings of language clear vision is impossible. You think you see, but are only running your finger through pubic hair. The rain was heavy enough to fall into this narrow street and pull shreds of cloud down with it. I expected the drops to strike my skin like a keyboard. But I only got wet. When there is no resonance, are you more likely to catch a cold? Maybe it was the uniform appearance of the drops which made their application to philosophy so difficult even though the street was full of reflection. In the same way, fainting can, as it approaches, slow the Yankee Doodle to a near loss of pitch. I watched the outline of the tower grow dim until it was only a word in my brain. That language can suggest a body where there is none. Or does a body always contain its own absence? The rain, I thought, ought to protect me against such arid speculations.

2 The body is useful. I can send it on errands while I stay in bed and pull the blue blanket up to my neck. Once I coaxed it to get married. It trembled and cried on the way to the altar, but then gently pushed the groom down to the floor and sat on him while the family crowded closer to get in on the excitement. The black and white flagstones seemed to be rocking, though more slowly than people could see, which made their gestures uncertain. Many of them slipped and lay down. Because they closed their eyes in the hope of opening their bodies I rekindled the attentions of love. High-tension wires very different from propensity and yet again from mirror images. Even if we could not remember the color of heat the dominant fuel would still consume us.

3 Androgynous instinct is one kind of complexity, another is, for example, a group of men crowding into a bar while their umbrellas protect them against the neon light falling. How bent their backs are, I thought. They know it is useless to look up— as if the dusk could balance both a glass and a horizon—or to wonder if the verb "to sleep" is active or passive. When a name has detached itself, its object, ungraspable like everyday life, spills over. A solution not ready to be taken home, splashing heat through our bodies and decimal points.

4 I tried to understand the mystery of names by staring into the mirror and repeating mine over and over. Or the word "me." As if one could come into language as into a room. Lost in the blank, my obsessive detachment spiraled out into the unusable space of infinity, indifferent nakedness. I sat down in it. No balcony for clearer view, but I could focus on the silvered lack of substance or the syllables that correspond to it because all resonance grows from consent to emptiness. But maybe, in my craving for hinges, I confused identity with someone else.

5 Way down the deserted street, I thought I saw a bus which, with luck, might get me out of this sentence which might go on forever, knotting phrase onto phrase with fire hydrants and parking meters, and still not take me to my language waiting, surely, around some corner. Though I am not certain what to expect. This time it might be Narragansett. Or black. A sidewalk is a narrow location in history, and no bright remarks can hold back the dark. In the same way, when a child throws her ball there is no winning or losing unless she can't remember her name because, although the street lamp has blushed on pink the dark sits on top of it like a tower and allows no more than a narrow cone of family resemblance.

6 I learned about communication by twisting my legs around yours as, in spinning a thought, we twist fiber on fiber. The strength of language does not reside in the fact that some one desire runs its whole length, but in the overlapping of many generations. Relationships form before they are written down just as grass bends before the wind, and now it is impossible to know which of us went toward the other, naked, unsteady, but, once lit, the unprepared fused with its afterimage like twenty stories of glass and steel on fire. Our lord of the mirror. I closed my eyes, afraid to resemble.

7

Is it possible to know where a word ends and my use of it begins? Or to locate the ledge of your promises to lean my head on? Even if I built a boundary out of five pounds of definition, it could not be called the shock of a wall. Nor the pain that follows. Dusk cast the houses in shadow, flattening their projections. Blurred edges, like memory or soul, an event you turn away from. Yet I also believe that a sharp picture is not always preferable. Even when people come in pairs, their private odds should be made the most of. You went in search of more restful altitudes, of ideally clear language. But the bridge that spans the mind-body gap enjoys gazing downstream. All this time I was holding my umbrella open.

8

I wondered if it was enough to reverse subject and object, or does it matter if the bow moves up or down the string. Blind possibility, say hunger, thickened. How high the sea of language runs. Its white sails, sexual, inviting to apply the picture, or black, mourning decline in navigation. I know, but cannot say, what a violin sounds like. Driftwood migrates toward the margin, the words gather momentum, wash back over their own sheets of insomnia. No harbor. No haul of silence.

9 There were no chronicles. The dimensions of emptiness instead of heroic feats. This was taken as proof that female means lack. As if my body were only layers on layers of window-pane. The whole idea of depth smells-fishy. But there are thicker transparencies where the sentence goes wrong as soon as you do something, because doing carries its own negative right into the center of the sun and blots out the metonymies of desire. In this neck of the womb. Later, sure enough, the applications fall away even if we cling to the exaggerated fireworks of lost purpose.

10 It is best to stop as soon as you hear a word in a language you don't know. Its opaqueness stands, not as a signpost to the adventures of misunderstanding, but a wall where touch goes deaf, and without explanations hanging in the air, waiting to be supported by the clotheslines of childhood. As I looked up, a boy approached me and offered to carry my bag because it was raining. Wet laundry flapped in the wind.

11
Heavy with soot, the rain drummed on the tin roof of the garage, eager to fall into language and be solved in the manner of mysteries. I tried to hear the line between the drumming and the duller thud on the street, like the phantom beat between two rhythms. An umbrella would have complicated the score. No gift of the singular: the sounds merged with traffic noise too gray to make a difference between woman and mother, or grammar and theology. Not like the children playing tag, throwing their slight weight into flight from the ever changing "it." Though the drops hit my face more gently than an investigating eye, the degree of slippage

12

Visibility was poor. The field limited by grammatical rules, the foghorns of language. On the sidewalk, people waiting for the bus looked out from under umbrellas and hoods, their eyes curtained by crosshatched rain, lids close to one another as when approaching sleep or pain. An adherence to darkness that refuses exact praise like reaching for a glass. Even when I had emptied mine, I had not gotten to the bottom of the things in plain view. A play of reflections and peculiar. The drops of water traveled diagonally across the paradigms as the bus moved on.

13 Because we cannot penetrate the soul, at most touch its outer lips with the reflected light of metaphor, the soul cannot know itself, but the dimmer light holds off loud breathing. It's not that our sense impressions lie, but that we understand their language. All through the linear seasons, the sun leaned on the shoulder of the road. Flocks of swallows lost vaster reasons to the sky. Salt travel. Statues which adorn the unconscious. My hopes crushed by knowledge of anatomy. Or is this another error, this theory of erosion, of all we cannot see?

14

On the fourth day, I took the rain in my mouth, and the fish sank deeper, lighting up in glints like time passing. The bus moved off, a long sliding door. Behind it, the row of houses suddenly larger, a mass of stone and wood to constrict your chest, as when you take a wrong turn to the side of your head which is dark with war and strangling and then are weak from loss of blood, a fishline wound around your neck. The dark was an obstacle. It would shortly come between me and the street, but its name made me want to touch its velvet beginning. Even though I've known complete rejection, words will still send me in pursuit of chimeras.

15 The room inside me has disappeared. At night, when all is quiet, I no longer hear the pictures shifting on the walls when I walk fast. Only the pump in the basement. I wonder whether the space has folded in on itself like a tautology, or been colonized. You think the wine has washed it out, and it's true that the mirror tilted at a reckless angle. I still have the floorplan with measurements, but now that nothing corresponds to it I can only take it as part of the emptiness I try to cover up with writing. To know by my blind spot. I have always wanted to dilate my landscape for the piano and the long labor of losing the self. Though I'm too nearsighted for clouds. If I had lived a different image.

16 If I were a mother I would naturally possess the pure crystalline logic which is the prerequisite and found in most well-appointed duals. Only, it is hard not to slip on the ice because there is no friction. Back to rougher sentences, I said, to the incomplete self and choice of desires. This postulated the vacancy as social rather than biological, and that it need not be filled. My body was calm, even naked, safe in its transparency.

17 This is where grammatical terror opens a distance between you and yourself in order to insert the mirror. And where you had hoped it would be a serene blue surface reflecting the flight of a bird or fancy, the waves rise up against each other and crash, strangling, screaming. What has become of logic? You know enough to skip explanation and displace your own weight in water. You hope the motion will wear itself out, its speed braked by words. History has taught you that all desires want to do away with themselves.

18 As long as I wanted to be a man I considered thought as a keen blade cutting through the uncertain brambles in my path. Later, I let it rust under the stairs. The image was useless, given the nature of my quest. Each day, I draw the distance to cover out of an anxiety as deep as the roots of language. I keep my eye on the compass while engaging the whole width of the field, and whereas, to others, I may look like a blur of speed from one point in time to another, I know I am not advancing an inch and will never arrive. Even if I could arrive the mirror would only show the other mirrors I have set up at every stop to catch the spirit of passage.

19

I worried about the previous occupants of the house, their traces burned off by daylight, unless a silence that grows more attentive as the hours wear on, some intangible bond among absent faces that outweighs my lack of purpose, the stillness of my posture. This is what makes the wallpaper so neutral as far as my affection goes. Its pattern rotates bits of landscape into orbits of definitive distance, like chrysanthemums celebrating All Souls'. It's not one of those pictures which I cannot get outside of because my language repeats it to me over and over, inexorably. From behind rocks, a boat is heading into a promise of archipelagos and coral islands, then travels into more faded patches as into so much fog.

20 With difficulty, the lamp outlined the limits of the dark where the objects sat so opaque, so at home in their silence. Not a clarity in which all likeness disappears, not breaking into space like a triumphant bird. Rather a dim lamp. Specks of dust moving in its cone, a fine attachment to light against the grain. As you might be stirred beyond all reason by the signs that form a word, easy motion, unbuttoned coincidence, clearing lips in a euphoria skidding toward the urban periphery. Or again, lost in the fog. It will cling to you no matter how heavy your boots.

21

The dusk swelled and lengthened every shadow, a version of time not suited to the ephemeral. For if you concern yourself with what belongs, but doesn't fit, turning your wrist like an angler out of season who, sure of his catch, throws his line into a cross section of air, the picture forces its particular application on you. Schools of thought emerge and are hooked in an instant. Others flit across the danger zone with tail fins flashing, and the alphabet ceases to be a disposition of the mind. I wondered, would a bracelet strengthen my analytical capacity.

22

As long as we can see the facts, you said, your eye frantic for touch, not a body of water which is beautiful in itself, but still a source of satisfaction in the stress of the air. So I tried to get the fact of rain down on the paper, the way it appeases the green, the haze rising too subtly to make me believe in an unbroken circle. It seeped away, into the pores, watering landscapes as removed from the pressure of perception as the nature of between, though the traces of ink looked a little like the shadows of clouds written on the ground. But any body being penetrated, all that is looked at

23 Given the distance of communication, I hope the words aren't idling on the map on my fingertips, but igniting wild acres within the probabilities of spelling. As a hawk describes circles whose inner emptiness bespeaks the power of gravity, where the lever catches on a cog of the world. There, the mild foreground for buying bread, for the averted doubt that the hand will encounter. There, with dizzy attention, I hold the because, another key to the bewitchment of words.

24

In the middle of rainy weather, sleep was pinning me down on the bed, lids barnacled shut with adjectives in color. Sleep, which cannot be divided from itself or into parts of speech, pushing a whole sea at my body so unable to swallow its grandiose and monotonous splendor. The air already slowing to the crucial stillness of noon. Would there ever again be ground for walking? I mean, the field of understanding does not extend to lying down. Later, writing would articulate the absence of voice, pictures, the absence of objects, clothes, the absence of body.

25 Never mind wholeness. Though it worries me that I have no sense of it. How can I approach the task without all of me gathered in one fist like, if I were a man I'd say, a stone I want to hurl or, as I'm not, an entire tune cradled on its fulcrum. Then I could be sure I won't get stuck if I try to sing. But I get stuck. Single notes occupy my voice with their vertical shadows, luminous blue, so long, so stretched, so content to sink slowly back into. Or a flock of swallows, alarmed suddenly, sucked crazily in all directions. The light appropriates, even to the unsounded spasms of treble and flight, and the fields stretch into what, lacking male parameters, must be nowhere.

26 It rained so much that I began to confuse puddles with the life of the mind. Perhaps what I had taken for reflection was only soaking up the world, a cross of sponge and good will through the center of the eye. But to describe the inner world, you know, by definition, even the patient definitions of psychology, is impossible. Hard to know if it can be lived. Revoked edge of water and dry land. A falling fear. The sudden color of a word. But it's the sky, pale gray, abundantly thrown back from far enough behind the eye, as you imagine an image, seeing earth in every direction.

27 The labyrinth of language. You know your way as you go in one bony side, but out the other you're lost in spiral frequencies, unsettled air as coiled as who'll believe it. Not for the first time has a wave broken on a hammer, anvil, or suddenly, which is a deep space where sentences breathe differently and the rider rocks in the stirrup. Caress of hair like dark approaching a naked voice. Or scales. If the air won't take my word I can't trace half a circle, falling, dizzy, into the confusion of canals.

28 If I promise day after day: tomorrow I'll come to see you, am I saying the same thing every day, or does a rainbow grow frenetic in the to and fro between eye and image, bits of light torn from a mirage which doesn't appease desire, but only fits into its own shape? Incestuous words, reflecting reference as mere decor or possibly a blanket. Orphaned so severely, the eye still trusts that emptiness is ready to receive the rain.

29 Once the word "pain" has replaced crying, behavior functions as landscape, and the philosopher can treat a question like an illness. The decisive moment is now, but dust has no particular object as it rises to the occasion, and only when I blink can I still see the distant shore. Nothing had prepared me for the end of monotony. I've always admired thin lines like the string of the marionette, which replaces consciousness catching in the hollow of the knee. But alone on a page. Or crossed out.

30

Look at that blue, you said, detaching the color from the sky as if it were a membrane. A mutilation you constantly sharpen your language for. I had wanted to begin slowly because, whether in the direction of silence or things have a way of happening, you must not watch as the devil picks your shadow off the ground. Nor the scar lines on your body. Raw sky. If everybody said, I know what pain is, could we not set clocks by the violent weather sweeping down from the north? Lesions of language. The strained conditions of colored ink. Or perhaps it is a misunderstanding to peel back skin in order to bare the mechanics of the mirage.

New Directions Paperbooks—A Partial Listing

For complete listing request free catalog from
New Directons, 80 Eighth Avenue, New York 10011   † Bilingual

For complete listing request free catalog from
New Directons, 80 Eighth Avenue, New York 10011   † Bilingual